Engineering for Inclusion

Held at the 2021 AIChE Annual Meeting

Boston, Massachusetts, USA and Online
7 - 11 November and 15 - 19 November 2021

ISBN: 978-1-7138-5685-6

Printed from e-media with permission by:

Curran Associates, Inc.
57 Morehouse Lane
Red Hook, NY 12571

CURRAN ASSOCIATES INC. proceedings.com

Some format issues inherent in the e-media version may also appear in this print version.

AIChE
120 Wall Street, FL 23
New York, NY 10005-4020

Phone: (800) 242-4363
Fax: (203) 775-5177

www.aiche.org

Additional copies of this publication are available from:

Curran Associates, Inc.
57 Morehouse Lane
Red Hook, NY 12571 USA
Phone: 845-758-0400
Fax: 845-758-2633
Email: curran@proceedings.com
Web: www.proceedings.com

TABLE OF CONTENTS

636455 Use of Environmental Sustainability Principles to Help Solve Environmental Problems of Concern Today

Abstract:

In 1987, the concept of sustainable development was first put forward in the Bruntland Report which was prepared for the World Commission on Environment and Development. The links between inequality, poverty, and environmental degradation were recognized; the report sought a way for a fairer world with less damage to the environment that supports communities. The report defined sustainable development as "Development that meets the needs of the present without compromising the ability of future generations to meet their own needs". As a consequence, environmental sustainability ensures that in meeting our needs for water, food, and shelter as well as engaging in activities that make our lives enjoyable, we don't damage the environment, deplete our natural resources, and protect global ecosystems. Important aspects tied in with address: (1) Behavioral economics; (2) Intergenerational equity; and (3) What to do with the developing world. As engineers, we will increasingly need to be considering these concepts.

Recent research activities aimed toward environmental sustainability have involved: condensate recovery in air handlers; improved building energy conservation; use of green roofs combined with cistern collection and rain gardens for storm water mitigation on campus buildings; modeling of pollutants fate and transport in the Nile River tributaries; use of electrochemical sensors in storm drains to characterize storm water runoff in urban roadways; use of vertical gardening systems and community gardens to address urban food deserts; lead contamination in drinking water systems physical/chemical treatment of hazardous wastes, particularly in the area of sonication (acoustic cavitation) and photocatalysts to treat organic contaminants; and waste plastics in the marine environment. Highlights of several of these research topics will be described in the context of environmental sustainability.

636472 Advancing Methodologies for Sustainability assessment of transition to Low Carbon and Circular Economy via integration of process systems engineering and macroeconomics

Abstract:

Decarbonization and implementing Circular Economy are two major goals for sustainable development to mitigate the challenges of climate change and increasing anthropogenic wastes. Among numerous pathways proposed for this transition, it is important to evaluate the systems level economic and environmental impacts of these pathways for selection of the most sustainable pathways. Industrial Ecology (IE) is a field of study that focuses on developing methodologies and computational tools to assess large scale impact of these interventions. One key method in IE utilizes the macroeconomic Input-Output model for evaluating the impact of planned transitions such as shifts to renewable energy, bio-based production etc. However, the existing techniques do not efficiently integrate process engineering and thermodynamics knowledge in sustainability assessments, thus limiting the automation in evaluation of new technologies under development at scale. Further, lack of engineering integration and reliance on only monetary based model introduces uncertainty in analysis due to price variation and obsolete models along with missing feedback in considering sustainability criteria in design. In this talk, I will present the advancement in systems level sustainability assessment techniques by addressing these limitations through integrating process systems engineering and thermodynamics with macroeconomic Input-Output framework. Specifically, I will present novel methodologies and computational tools that will enable fast, and reproducible quantitative evaluation of systems level environmental and economic impacts of adopting emerging technologies for decarbonization and circular economy pathways.

637601 Bioconversion of plastic wastes into value-added products using thermal oxo-degradation

Abstract:

Over 8 billion tons of plastics have been produced, and the available strategies for its disposal have proven insufficient due to economic and environmental sustainability limitations. As a result, most of all plastic ever made has ended or will end in landfills and our ecosystems. Plastics are not biodegradable because the breakdown of the polymers into molecules that microorganisms can utilize is estimated to take thousands of years. However, combining thermal oxo-degradation (TOD) with biocatalytic conversion can drastically reduce the rate-limiting step of plastic degradation and open an avenue for plastic waste upgrading. TOD rapidly deconstructs polymer backbones while adding oxygen functionalities to the fragmented products. The products of this process with plastic feedstocks are mixtures of fatty substrates - including fatty acids and fatty alcohols – and hydrocarbons. In this study, a screening process was carried out to select a suitable microorganism for the bioconversion of thermally oxo-degraded high-density polyethylene (HDPE), the world's most common plastic. The TOD products were hydrophobic and not emulsified before testing carbon utilization, resulting in a solid floating at the surface of the culture. Remarkably, the selected organism overcame mass transfer limitations in the liquid culture and grew with TOD products as its sole carbon source. The growth profile was even comparable to glucose controls. This microorganism could be leveraged as a microbial cell factory, or its biomass could be turned into single cell protein, resulting in waste upcycling. These findings serve as proof-of-concept and showcase the potential of our novel approach.

637602 Kasim A. Adewuyi Poster

Abstract:

This poster presents a new optimal charging methodology called Constant Gradient Constant Voltage (CGCV), which limits mechanical crack propagation at the surface of the particles in electrodes by regulating the current to maintain a constant lithium gradient at the surface. One important cause of battery degradation is the stress that is developed due to the concentration gradient inside active particles. The proposed new charging protocol actively controls the concentration gradient by utilizing the fact that the lithium diffusivity is strongly state-of-charge (SOC) dependent. Here, the diffusivity is not on-time measurable, so a modeling-based approach is incorporated, in which the charging protocol is guided based on the concentration gradient predicted from the modeling results. The upper limit of the concentration gradient is defined based on the maximum stress during the conventional Constant Current Constant Voltage (CCCV) charging. Then, two values (72.42% and 57.26% of the maximum gradient in CCCV for high and low CGCV, respectively) are considered for the demonstration of the proposed algorithm. The experimental results under three charging scenarios of high CGCV, low CGCV, and standard CCCV show similar degradation rates with a 25% or 17% reduction in charging time from the two CGCV methods compared to CCCV charging. This is a new protocol considering battery physics on particle level and has great potential to address the key challenge of regulating battery degradation during fast charging of Electric Vehicles (EVs).

637603 Impact of Crowding Agent on Supramolecular Complexation

Abstract:

Supramolecules such as, octa-acid (OA), tetra-endo-methyl octa-acid (TEMOA) and Positand are host molecules that possesses a water-soluble outer coat and a deep hydrophobic pocket suitable for binding a range of non-polar guests in aqueous solution. Via the hydrophobic effect, OA readily assembles into dimeric capsules in the presence of suitably sized guests. However, experiments have shown that the Positand does not assemble into complexes as easily as OA. In studies involving macromolecular crowding, effects that have been observed include formation of macromolecular complexes in solution and the stabilization of folded proteins and protein/ligand complexes. The magnitude of the effect has been shown to strongly depend upon the relative sizes and shapes of concentrated crowding species. Here, we study via molecular simulation techniques, the impact of molecular crowders on the assembly properties of supramolecules. Our results so far shows that by varying the concentration and/ or molecular weight of the crowding agent molecule in the simulation system, relatively improved complex stabilization is achieved.

637604 Thermal Properties of Heat-treated Coal Under Inert Conditions

Abstract:

Research and development into coal for building/construction applications is becoming more popular e.g., coal as a filler in polymer materials [1]. These alternative applications will involve developing new processes which require simulations to predict process economics at scale based on mass and energy balances. For any coal processing method, understanding coal properties is essential and so far, numerous experimental studies offer information on the thermal behavior with proposed heating value and density correlations [2]. However, most of this literature is limiting because it provides information on the thermal properties of coal and not on coal derivatives formed via heat treatment. In any coal heat treatment process, understanding the thermal characteristics of coal-derived materials would offer valuable insight in assessing process energetics. The purpose of this study is to characterize the thermal behavior of bituminous coal undergoing thermal treatment and to simulate this behavior using Aspen Plus and the Kirov correlation for predicting heat capacity [3]. This study also aims to effectively propose a model than can accurately describe the heat capacity of the coal-based material as it as it becomes more vitreous in nature. If successful, the model would serve as a framework for simulating processes that yield coal-derived alternatives and a foundation for scaling up these processes.

637605 Suraj Aryan Poster

Abstract:

The insulation in buildings is very important. Insulation used in the building can be classified by material: Organic and Inorganic insulation. Organic materials made of Styrofoam or polyurethane are at high risk for fire, less than 5 seconds during heating and the average time to spread flame is 50 seconds. On the other hand, inorganic insulation such as mineral wool and glass wool has an excellent fire resistance characteristic but its absorbability is very high, so there is a relapse in its insulation performance. The basic parameter that includes thermal coatings is the coefficient of thermal conductivity λ. Its value depends on the temperature, the density, the type of pores, the type of gas contained in the cells and the moisture content. Inorganic insulation materials such as glass fiber or mineral wool are extremely vulnerable to moisture so that they absorbed water 4~8% of their weight. And we know the value of the thermal conductivity coefficient rises as the water content of the material increases. While the thermal conductivity of air is 0.026 W/mk, water has 0.598W/mk which is 23 times the thermal conductivity of air. Its deep penetration into the material causes displacement of air, which results in an increase in the insulation's thermal conductivity coefficient, because of that it is difficult to expect a proper performance as insulation in a high humidity area.

637606 Preliminary risk assessment of traffic accidents in Colombia involving loss of containment of liquid hydrocarbons

Abstract:

In Colombia, the environmental regulatory authorities have recently implemented standard procedures for the transportation of liquid fuels, requiring a mandatory quantitative assessment of the associated risks of such activities. This assessment must involve computer modelling of storage leakage or spill scenarios, and their corresponding impacts (e.g. fire, explosion, toxic release) in order to develop and design suitable contingency plans. This implies an understanding of potential hazards, the identification of the source of failure, quantification of probability of occurrence, and the assessment of potential impacts. The environmental authority must be given enough information to permit the transportation of a specific hazardous material, within the planned route, in the specific vehicle, at a certain date. As this is a new regulation, and most transportation companies cannot afford the use of licenced software tools or to hire
specialized companies to do the assessment, there is need to incorporate free software tools to accomplish such task.

In this regard, this study was focused on identifying the most typical petrochemical materials transported in the Colombian territory and their corresponding volumes. Then, based upon a multi-dimension decision method, most hazardous materials were identified. This information was used to identify two cases of study to do the modelling of loss of containment incidents and to assess their potential impacts. The geographic location for the modelling was defined considering the historical reports of incidents involving transportation of liquid fuels. Modelling was carried out with ALOHA by using surrogate molecules that were selected based upon a comprehensive properties' analysis.

637607 Hierarchical Zeolite Composite Catalyst from the Aerosol-assisted Encapsulation of ZSM-5in Ordered MCM-41 Microspheres

Abstract:

Despite the high activity of zeolites for the cracking of hydrocarbon molecules, its rapid deactivation due to coke deposition remains a drawback. These microporous zeolites also need to be embedded in meso/macro porous supports to improve heat and mass transport within a fixed bed reactor as well as reduce the pressure drop across the reactor. By dispersing zeolites within the matrix of mesoporous silica such as MCM-41, the MCM-41 can serve as a protective coating by acting as a sacrificial material for coke deposition and will also enable improved transfer of reactants to the zeolites. We report new findings in the use of an aerosol-assisted synthesis technique to incorporate ZSM-5 zeolites into the matrix of spherical MCM-41 nanoparticles. In this work, pre-synthesized ZSM-5 was introduced into silica precursor solution and aerosolized through a furnace using N2 as carrier gas. Detailed characterization of the synthesized catalysts using electron microscopy (SEM and TEM), BET analysis and X-ray diffraction (XRD) confirmed the encapsulation of the ZSM-5 within the MCM-41 matrix. To determine the accessibility of reactants to the encapsulated ZSM-5 through the pores of MCM-41, we performed the catalytic reduction of 4-nitrophenol to 4-aminophenol over encapsulated Ni/ZSM-5. We have also shown the efficacy of this composite catalyst towards improving the catalytic pyrolysis and upgrading of miscanthus x gigantus biomass to liquid fuels. The mesoporous matrix of the MCM-41 can also serve as support for metal catalysts leading to the coupling of multi-stage cracking and upgrading reactions into one catalytic system. These properties have immense implications for hydrocarbon processing and the production of sustainable fuels.

637608 Metabolic Modeling of *Pseudomonas Putida* KT2440 to Understand and Improve the Breakdown of Plastic Waste

Abstract:

Polyethylene terephthalate (PET), the polymer used to manufacture plastic beverage bottles, is the world's highest production-volume synthetic plastic. While synthetic polymers are typically resistant to biodegradation, organisms have been discovered that can decompose PET into soluble but toxic monomers, ethylene glycol (EG) and terephthalic acid (TPA). *Pseudomonas putida* is a soil bacterium known for its receptiveness to genetic engineering and ability to metabolize a wide assortment of organic compounds and pollutants. We seek to understand the pathways by which *P. putida* can metabolize EG and/or TPA and whether its growth on these compounds could be improved by genetic modification. Our analyses are based on whole-genome metabolic models of *P. putida* accessed with the web platform KBase and locally implemented in a Python programming environment, COBRApy. These models are used to generate flux balance analysis (FBA) outputs that allow us to quantitatively predict the organism's growth in different media. FBA results indicate that *P. putida* KT2440, described by the model iJN1463, is capable of biomass formation on EG. We are currently investigating growth on TPA. Our research has uncovered metabolic "bottlenecks" that, if alleviated by genetic engineering, could improve the ability of *P. putida* to metabolize these carbon sources.

637611 The Pathophysiological Effects of Fluid-Structure Interaction of Species Transported and Transformed from Ambient to Human Respiratory System

Abstract:

The physiological system in humans consist of several organs that are structured to relate chemically and physically with complex fluids which aid to carry different nutrients and gaseous species (oxygen) necessary for proper functioning of the human system. However, the quality of the ambient atmosphere and its physiological end effect plays a tremendous role in determining homeostasis (stability) of the human body systems. For instance, anthropological alteration of the quality of air such as incomplete combustion of hydrocarbon can lead to release of injurious gases such as carbon monoxide and if inhale by human can distort the composition of blood leading to the formation carboxyhemoglobin. This physiological distortion is fatal in its consequence. Furthermore, with the increased interest in correlating ambient fluid quality with human pathophysiological conditions, it has become immensely significant to understand the mechanism of species transported and transformed via the respiratory tract, the major gateway to human complex systems. Thus, the research is aimed at studying the pathophysiological impact of ambient species on human physiological system through detail analysis of the flow process of species and its mechanism of interaction as they are transported from the bulk ambient fluid to the respiratory organs. A detail qualitative study will be conducted to gather data on the composition of the ambient atmospheric fluids at different geographical locations and coupled these data into modelling platform designed to study the mechanism of transportation and transformation of species as they interact with the organs of the respiratory system. The goal of this work is to develop a prognosticative and diagnostic modelling platform that can successfully analyze and predict the pathophysiological effects of ambient fluid on human systems while offering suggestions on range of preventive and therapeutic measures to correct these physiological anomalies.

637613 Bianca Mitchell

Abstract:

The use of cellulose nanocrystals (CNCs) has seen an increase in interest for numerous industrial applications due to their low cost, high abundance, and lightweight features. Due to the extremely hydrophilic nature of CNCs, the number of matrices that the particle can be dispersed in is limited. Therefore, functionalizing the surface groups of the crystal is necessary in to increase its hydrophobicity and allow for dispersion in commonly used hydrophobic polymer matrices. Following the modification of the particle, it also critical to understand how the microstructure of the composite changes during post-processing. In-situ cure monitoring can reveal vital information regarding the microstructure properties of the composites, which is directly related to the overall performance of the composite (tensile strength).

In this work, we explore the dynamics of a surface-functionalized CNC during the curing in a commonly used thermosetting resin (EPON 828). X-ray photon correlation spectroscopy (XPCS), coupled with differential scanning calorimetry (DSC), reveal drastic changes in the dynamics and curing kinetics of the composite as a function surface modification. To confirm that the crystals were successfully functionalized, both Fourier transform infrared, and x-ray diffraction was utilized to verify results before dispersing the crystals in the hydrophobic thermosetting epoxy resin. Two levels of modification (7 & 10 %) were chosen to perform observations at a constant loading of 5 wt%. The goal of this study is to compare the macroscale curing dynamics using DSC to non-isothermally cure with the microscale dynamics using XPCS. By using XPCS to compliment the macroscale results obtained from DSC, it is possible to fully understand the preliminary and final dynamics displayed by the composite on a microscale level that cannot be observed using simply DSC. With the results we've obtained using both methods, the degree of cure using either an isothermal or non-isothermal procedure can be better understood, and techniques can be tailored to achieve desired properties.

637019 From Pledges to Practice: Turning Commitments into Action to Create More Inclusive Institutions

Abstract:

The United States and the world are in the midst of a watershed moment of racial reckoning. In the wake of global protests for racial justice, billions have been pledged, and thousands of companies, universities, and associations have made commitments to dismantle systemic racism and work toward social justice. There's a reason. For institutions to thrive, they must foster an inclusive climate, one in which every stakeholder feels that they can bring their whole selves into the classroom, laboratory and the workplace. And yet, fostering true systems change has proven elusive.

Engineers are astute at tackling complex problems. Who, then, are better equipped to grapple with the complexity of increasing diversity, and fostering equity and inclusion (DEI) in higher education, industry and other domains at scale?

This talk synthesizes the latest research on diversity into a practical value proposition, through the lens of an engineer. It makes the case for why diversity and inclusion matter, and offers a systems approach to creating more equitable and inclusive workplaces and campuses.

639806 Academic Ableism and its Alternatives

Abstract:

About the speaker: Dr. Jay T. Dolmage is a professor of English language and literature at the University of Waterloo. He is committed to disability rights in his scholarship, service, and teaching. His work brings together rhetoric, writing, disability studies, and critical pedagogy. He is the author of *Academic Ableism: Disability and Higher Education* (2017) and *Disability Rhetoric* (2014). He is also the founding editor of the *Canadian Journal of Disability Studies*.

639807 Increasing Participation of Students with Disabilities in STEM

Abstract:

Students with disabilities are a significantly underrepresented demographic in the STEM disciplines. To help alleviate this inequality, we, at the University of Delaware, have developed the unique summer outreach program "Research Experience for Undergraduates Science and Engineering Leadership Initiative." This innovative, NSF-funded program encourages undergraduates with physical, psychological, or learning disabilities to take an active role in the scientific endeavors at a research-intensive University. This integrative measure is designed to empower the participants to pursue research-related careers and professions related to STEM disciplines. In our presentation, we will not only discuss the details of our program but aim to give experience-based, practical advice on how to best accommodate students with disabilities in research laboratories and ways to implement similar programs in other institutions. We will discuss the advocacy work of the American Chemical Society Committee on Chemists with Disabilities.

About the speaker: Dr. Sharon Rozovsky is a professor of biochemistry at the University of Delaware. She is one of the faculty directors of the Science & Engineering Leadership Initiative, a research experiences program designed for the advancement of students with disabilities who are interested in careers in chemistry, biochemistry, or aligned fields. She served on the Committee on Chemists with Disabilities of the American Chemical Society in 2018–2020.

637411 The Hard Skills of Emotional Intelligence for Effective Engineering Leaders

Abstract:

High performing teams and leaders with emotional intelligence are more likely to achieve their goals. This workshop featuring Dr David Caruso of Yale's Center for Emotional Intelligence, focuses on the ability model of emotional intelligence and begins with a hands-on illustration of the four “hard” skills of EI, followed by demonstrations of the notion that “emotions are data” and the cognitive costs of emotion suppression. After a review of a brief case study, we then provide participants with a means to apply these skills to situations faced by leaders in engineering using the Emotional Intelligence Blueprint.

AUTHOR INDEX

AIChE
120 Wall Street, FL 23
New York, NY 10005-4020

ISBN 978-1-7138-5685-6

CURRAN ASSOCIATES INC.
proceedings.com